Physics Foundations

Your First Guide to the Universe

Table of Contents

1. Introduction
2. Chapter 1: Understanding the Basics of Physics
3. Chapter 2: Core Concepts in Physics
4. Chapter 3: Essential Tools and Resources for Physics
5. Chapter 4: Mechanics
6. Chapter 5: Thermodynamics
7. Chapter 6: Electromagnetism
8. Chapter 7: Waves and Optics
9. Chapter 8: Modern Physics
10. Chapter 9: Conducting Simple Physics Experiments
11. Chapter 10: Applying Physics in Real Life
12. Conclusion
13. Appendices

This Table of Contents offers a logical progression through the basics of physics, ensuring readers build a strong foundation before exploring more complex concepts and applications.

Book outline

Introduction

Purpose of this first Guide: Explain the goal of providing a beginner-friendly overview of fundamental physics concepts and applications.

Who This first Guide is For: Students, enthusiasts, or anyone new to physics.

Chapter 1: Understanding the Basics of Physics

1.1 What is Physics?: Overview of physics as the study of matter, energy, and fundamental forces.

1.2 Why Study Physics?

1.3 The Scientific Method in Physics

1.4 Key Concepts and Terminology

1.5 SI Units and Measurement: Importance of SI units, commonly used units in physics (meter, kilogram, second, etc.).

1.6 Branches of Physics

- Mechanics
- Electromagnetism
- Thermodynamics
- Quantum Mechanics
- Relativity

1.7 Fundamental Laws and Principles

1.8 Measurement and Observation

1.9 Conclusion

Chapter 2: Core Concepts in Physics

2.1 Motion and Kinematics:

- Definition and types of motion (linear, rotational, etc.).
- Key quantities: speed, velocity, acceleration.
- Equations of motion.

2.2 Forces and Newton's Laws:

- Types of forces: gravitational, normal, friction, tension.
- Newton's Three Laws of Motion with examples.

2.3 Energy and Work:

- Forms of energy: kinetic, potential, thermal, and others.
- Work-energy theorem.
- Conservation of energy.

2.4 Momentum:

- Definition of momentum and impulse.
- Conservation of momentum in collisions.

2.5 Gravity:

- Universal law of gravitation.
- Weight and Gravitational Acceleration

2.6 Waves and Oscillations

- Type of Waves
- Properties of Waves

2.7 Electric and Magnetic Fields

- Electric Fields
- Magnetic Fields

2.8 Summary

Chapter 3: Essential Tools and Resources for Physics

3.1 Laboratory Equipment and Measurement Tools

- Basic Measurement Instruments
- Force and Motion Measurement
- Electrical Measurement Instruments
- Optics and Light Measurement
- Temperature Measurement

3.2 Mathematics in Physics:

- Algebra, trigonometry
- Calculus
- Vectors and scalars and geometry
- Graphs and Data Analysis

3.3 Software and Simulations:

- Data Collection and Analysis Software
- Simulation Software
- Circuit Design Software

3.4 Online Resources for Physics Learning

- Educational Platforms
- Video Tutorials and Channels
- Interactive and Virtual Labs

3.5 Reference Books and Textbooks

3.6 Apps and Mobile Tools

3.7 Conclusion

Chapter 4: Mechanics

4.1 What is Mechanics?

4.2 Types of Motion

- Linear Motion
- Rotational Motion
- Projectile Motion

4.3 Energy and Work in Mechanics

- Work
- Kinetic Energy
- Potential Energy
- Conservation of Energy

4.4 Rotational Motion and Torque

- Torque
- Motion and inertia
- Angular Velocity and Acceleration

4.5 Applications of Mechanics

4.6 Summary

Chapter 5: Thermodynamics

5.1 What is Thermodynamics?

5.2 Key Concepts in Thermodynamics

- System and Surroundings
- Heat and Temperature
- Work(W)
- Internal Energy (U)

5.3 Laws of Thermodynamics:

- The Zeroth Law of Thermodynamics
- The First Law of Thermodynamics (Conservation of Energy)
- The Second Law of Thermodynamics (Entropy)
- The Third Law of Thermodynamics (Absolute Zero)

5.4 Thermodynamic Processes

- Isothermal Process
- Adiabatic Process
- Isobaric Process
- Isochoric Process (Isovolumetric)

5.5 Understanding Entropy and the Arrow of Time

5.6 Applications:

- Heat Engines
- Refrigerators and Heat Pumps
- Biological Systems
- Environmental Science and Climate

5.7 Summary

Chapter 6: Electromagnetism

6.1 What is Electromagnetism?

6.2 Key Concepts in Electromagnetism

- Electric Charge
- Electric Fields (E-Fields)
- Magnetic Fields (B-Fields)
- Electromagnetic Force

6.3 Maxwell's Equations

6.4 Electromagnetic Induction

6.5 Electromagnetic Waves

6.6 Practical Applications of Electromagnetism

- Electric Motors
- Generators
- Transformers
- Wireless Communication
- Medical Imaging

6.7 Summary

Chapter 7: Waves and Optics

7.1 What is a Wave?

7.2 Properties of Waves

- Wavelength (λ)
- Frequency (f)
- Amplitude (A)
- Wave Speed (v)
- Phase

7.3 Sound Waves

7.4 Light as an Electromagnetic Wave

7.5 Basic Concepts of Optics

- Reflection
- Refraction
- Diffraction
- Interference

7.6 Optical Instruments

- Microscopes
- Telescopes
- Cameras
- Lasers

7.7 The Electromagnetic Spectrum

7.8 Practical Applications of Waves and Optics

7.9 Summary

Chapter 8: Introduction to Modern Physics

8.1 What is Modern Physics?

8.2 Key Concepts in Modern Physics

- Quantum Mechanics
- Theory of Relativity

8.3 Atomic and Nuclear Physics

- Atomic Structure
- Nuclear Physics

8.4 The Standard Model of Particle Physics

8.5 Key Experiments in Modern Physics

- Double-Slit Experiment
- Photoelectric Effect
- Large Hadron Collider (LHC)

8.6 Modern Physics and Technology

8.7 Philosophical Implications of Modern Physics

8.8 Summary

Chapter 9: Conducting Simple Physics Experiments

DIY Experiments:

9.1 Safety Guidelines

9.2 Mechanics Experiments

- Measuring Acceleration Due to Gravity
- Building a Simple Pendulum

9.3 Thermodynamics Experiments

- Observing Heat Transfer
- Creating a Homemade Thermometer

9.4 Electromagnetism Experiments

- Making a Simple Electromagnet
- Generating Electricity with a Lemon

9.5 Optics Experiments

- Observing Refraction with a Glass of Water
- Creating a Simple Lens

9.6 Waves Experiments

- Observing Sound Waves
- Creating a Ripple

9.7 Summary

Chapter 10: Applying Physics in Real Life

10.1 Physics in Transportation

- The Science of Motion
- Safety Features

10.2 Physics in Communication Technology

- Wireless Communication
- Sound and Microphones

10.3 Physics in Healthcare

- Medical Imaging
- Lasers in Medicine

10.4 Physics in Energy and Power

- Electricity in Daily Life
- Renewable Energy Sources

10.5 Physics in Everyday Gadgets

- Smartphones
- Refrigerators and Air Conditioners
- LED Lights

10.6 Physics in Sports and Fitness

- Motion and Force in sports
- Wearable Technology

10.7 Physics in Household Activities

- Cooking
- Washing Machines

10.8 Physics and the Environment

- Understanding Weather
- Climate Change

10.9 Physics and Human Movement

- Walking and Running
- Hearing and Vision

10.10 Summary

Conclusion

Summary of Key Concepts: Recap of the main topics covered in this first guide.

Next Steps for Further Study: Recommendations for more in-depth study resources.

Encouragement for Curiosity and Experimentation: Inspiring readers to continue learning and exploring physics.

Appendices

Formulas and Constants: Handy list of essential physics formulas and universal constants.

Glossary of Physics Terms: Definitions of commonly used physics terms.

Acknowledgments

This book outline provides a structured approach to learning physics, guiding readers from foundational concepts to hands-on experiments and real-world applications. Each chapter can be further expanded to include examples, exercises, and visual aids, helping readers engage with and apply physics concepts.

----

1. Introduction

Welcome To The "**Physics Foundations: Your First Guide to the Universe,**" your gateway to understanding the fundamental principles that shape our universe. Whether you're a curious beginner, a student looking for clear explanations, or simply intrigued by the mysteries of matter and energy, this first guide is designed to make physics approachable and engaging.

Physics is more than just equations and theories; it's the study of the natural laws that govern everything from the smallest particles to the largest galaxies. It reveals why objects fall, how electricity powers our homes, and even how time and space are interconnected. By learning physics, you're not only expanding your knowledge but also empowering yourself to see the world with fresh eyes and a deeper sense of wonder.

In this **Physics Foundations: Your First Guide to the Universe,** we'll walk you through essential concepts, introduce you to key terms, and explain the foundational laws that scientists and engineers rely on every day. You'll gain hands-on experience with simple experiments, learn practical problem-solving strategies, and explore how physics applies to real-world technology and innovation.

Ready to dive in? Let's unlock the secrets of the universe, one principle at a time.

Chapter 1: The Basics of Physics

Physics is the science of understanding the fundamental nature of everything in our universe. It explains how objects move, interact, and transform. Through physics, we gain insights into the forces that shape our world, from the smallest particles to the vastness of space. This chapter will introduce the essential principles, units, and branches of physics to set the stage for deeper exploration.

1.1 What is Physics?

Physics is a branch of science that studies matter, energy, and their interactions. It seeks to answer questions like, "Why do objects fall?" "How does light travel?" and "What causes stars to shine?" Physics serves as the foundation for other sciences, including chemistry, biology, and engineering, by providing essential principles that explain how the universe operates.

1.2 Why Study Physics?

Understanding physics is crucial for anyone interested in science or technology. Physics not only fuels scientific innovation, but it also enhances our understanding of natural phenomena. Whether you're aiming to pursue a career in science or simply curious, physics enables you to think critically, solve complex problems, and see the world from a new perspective.

1.3 The Scientific Method in Physics

Physics relies on the scientific method, a systematic process for exploring and understanding phenomena. Here's a quick breakdown:

Observation: Identifying and recording a phenomenon.

Hypothesis: Proposing an explanation based on current knowledge.

Experiment: Testing the hypothesis through controlled experiments.

Analysis: Reviewing data to confirm or refute the hypothesis.

Conclusion: Drawing conclusions and refining theories or proposing new hypotheses.

This method allows physicists to investigate questions rigorously, forming the backbone of scientific discovery.

1.4 Key Concepts and Terminology

To build a strong foundation, let's review some fundamental physics concepts and terms.

1.4.1 Matter and Energy

Matter: The "stuff" that makes up the universe, from atoms and molecules to stars and galaxies. Matter has mass and occupies space.

Energy: The ability to do work or produce change. Energy exists in various forms, including kinetic, potential, thermal, and electromagnetic energy.

1.4.2 Force and Motion

Force: A push or pull on an object that causes it to move, stop, or change direction. Newton's laws of motion describe how forces affect objects.

Motion: The change in an object's position over time. Motion is characterized by speed, velocity, and acceleration.

1.4.3 Space and Time

Space: The three-dimensional expanse in which objects and events occur. Physics examines how objects are positioned and interact within space.

Time: The ongoing progression of events. Time helps us understand sequences and durations, essential for studying motion and change.

1.5 Units and Measurement

To make accurate observations, physics relies on standardized units. In physics, we use the **International System of Units (SI):**

Quantity	SI Unit	Symbol
Length	meter	m
Mass	kilogram	kg
Time	second	s
Temperature	kelvin	K
Electric current	ampere	A
Amount of substance	mole	mol
Luminous intensity	candela	cd

Understanding these units is essential, as they allow us to measure, compare, and predict outcomes consistently.

1.5.1 Converting Units

Physics often involves converting between units, such as meters to kilometers or seconds to minutes. Knowing how to convert units is fundamental in solving physics problems. For example:

1 kilometer (km) = 1000 meters (m)

1 hour (h) = 3600 seconds (s)

1.6 Branches of Physics

Physics encompasses a wide array of subfields, each focusing on different aspects of the natural world:

1.6.1 Classical Mechanics

Description: Studies the motion of objects and the forces acting upon them. This branch includes Newtonian mechanics and covers everything from a falling apple to planetary motion.

Key Concepts: Newton's laws of motion, kinematics, and dynamics.

1.6.2 Electromagnetism

Description: Examines electric and magnetic fields and their interactions. Electromagnetism is the basis for technologies like electric motors, generators, and electronics.

Key Concepts: Electric charge, current, voltage, magnetism, and electromagnetic waves.

1.6.3 Thermodynamics

Description: Focuses on heat, temperature, and energy transfer. Thermodynamics is fundamental to understanding engines, refrigerators, and weather patterns.

Key Concepts: Laws of thermodynamics, heat transfer, and entropy.

1.6.4 Quantum Mechanics

Description: Deals with the behavior of particles on atomic and subatomic scales. Quantum mechanics explains phenomena that classical mechanics cannot, like the behavior of electrons in atoms.

Key Concepts: Wave-particle duality, uncertainty principle, and quantum states.

1.6.5 Relativity

Description: Introduced by Albert Einstein, relativity describes how objects behave at high speeds and in strong gravitational fields. It revolutionized our understanding of time and space.

Key Concepts: Special relativity (time dilation and length contraction) and general relativity (gravity as curvature of space-time).

1.7 Fundamental Laws and Principles

Physics is governed by certain fundamental laws that apply universally. Here are a few key principles:

1.7.1 Newton's Laws of Motion

First Law (Inertia): An object at rest will stay at rest, and an object in motion will stay in motion, unless acted upon by a force.

Second Law (Force): The force acting on an object is equal to its mass times its acceleration, $F = ma$.

Third Law (Action-Reaction): For every action, there is an equal and opposite reaction.

1.7.2 Conservation Laws

Conservation of Energy: Energy cannot be created or destroyed, only transformed from one form to another.

Conservation of Momentum: In a closed system, the total momentum before an event is equal to the total momentum after.

Conservation of Charge: Electric charge is conserved in isolated systems.

1.8 Measurement and Observation

In physics, making accurate measurements and observations is critical:

Precision vs. Accuracy: Precision refers to the consistency of measurements, while accuracy is how close a measurement is to the true value.

Error and Uncertainty: All measurements come with some degree of uncertainty. Understanding sources of error is essential for interpreting data accurately.

Practical Tips for Measurement

Use the Right Tools: Choose measuring tools suited for the quantity you're observing (e.g., rulers for length, stopwatches for time).

Record Carefully: Note units and be mindful of rounding errors.

Repeat Measurements: Repeating measurements and averaging results improves accuracy.

1.9 Conclusion

In this chapter, we've explored the basics of physics, including fundamental concepts, SI units, and key branches. By understanding these essentials, you're well-prepared to dive deeper into specific areas of physics. Remember, physics is not just a study of theoretical concepts but a way of describing and understanding the real world.

The following chapters will build on this foundation, introducing you to mechanics, electromagnetism, thermodynamics, and more. So, let's continue the journey into the fascinating world of physics!

Chapter 2: Core Concepts in Physics

In this chapter, we'll dive into the core principles that form the foundation of physics. These concepts are universal, explaining how objects move, how forces act, and how energy transforms. By understanding these basics, you'll gain a solid foundation for exploring the more complex ideas introduced in later chapters.

2.1 Motion and Kinematics

Kinematics is the study of motion without considering the forces that cause it. It helps us describe how objects move over time and space.

2.1.1 Types of Motion

Linear Motion: Movement in a straight line (e.g., a car moving down a road).

Projectile Motion: The motion of an object thrown into the air, affected by gravity (e.g., a soccer ball kicked into the air).

Rotational Motion: Movement around a central point (e.g., Earth's rotation around its axis).

2.1.2 Key Quantities in Kinematics

Distance and Displacement: Distance is the total path covered, while displacement is the shortest straight-line distance from start to finish, with direction.

Speed and Velocity: Speed is how fast an object is moving, while velocity is speed with direction.

Acceleration: The rate of change of velocity over time. Positive acceleration means increasing velocity, while negative acceleration (deceleration) indicates decreasing velocity.

2.1.3 Equations of Motion

In uniformly accelerated motion, we can use specific equations to describe the relationship between displacement, initial velocity, final velocity, acceleration, and time. Commonly used kinematic equations include:

$v = u + at$

$s = ut + \frac{1}{2}at^2$

$v^2 = u^2 + 2as$

Where:

v = final velocity

u = initial velocity

a = accelerationn

t = time

s = displacement

2.2 Forces and Newton's Laws of Motion

Forces are pushes or pulls that act upon objects, influencing their motion. The study of forces and their effects is known as dynamics.

2.2.1 Types of Forces

Gravitational Force: The attractive force between two masses. On Earth, this force gives objects weight.

Normal Force: The support force exerted by a surface on an object resting on it.

Frictional Force: The force that opposes the motion of an object across a surface.

Tension Force: The force transmitted through a string, rope, or cable when it is pulled tight.

Applied Force: A force applied directly to an object by another object (e.g., pushing a box).

2.2.2 Newton's Three Laws of Motion

First Law (Law of Inertia): An object at rest will stay at rest, and an object in motion will stay in motion unless acted upon by an external force.

Second Law (Force and Acceleration): The acceleration of an object depends on the net force acting on it and its mass, represented by .

Third Law (Action and Reaction): For every action, there is an equal and opposite reaction.

2.3 Energy and Work

Energy is the capacity to do work, and it exists in various forms, such as kinetic, potential, and thermal energy.

2.3.1 Types of Energy

Kinetic Energy: Energy of motion, given by the formula $KE = \frac{1}{2}mv^2$

Potential Energy: Stored energy based on an object's position or configuration. The most common form is gravitational potential energy, $PE = mgh$, where g is the gravitational acceleration.

Thermal Energy: Energy related to temperature, often resulting from the kinetic energy of particles in an object.

Chemical Energy: Stored in chemical bonds and released during chemical reactions.

2.3.2 Work and Power

Work: The energy transferred by a force acting over a distance. Defined as $W = F.d.\cos(\theta)$, Where:

W = Work

F = force

d = displacement

θ = angle between force and displacement direction

Power: The rate at which work is done, given by $P = \frac{W}{t}$, where:

P = power

W = work

t = time

2.3.3 Conservation of Energy

Energy cannot be created or destroyed, only converted from one form to another. This is known as the **Law of Conservation of Energy**. For example, a roller coaster at its highest point has maximum potential energy, which transforms into kinetic energy as it descends.

2.4 Momentum and Impulse

Momentum is a measure of an object's motion, combining its mass and velocity. Impulse describes the change in momentum caused by a force.

2.4.1 Momentum

Definition: Momentum (p) is the product of mass and velocity: $p = mv$.

Conservation of Momentum: In a closed system with no external forces, the total momentum before an event is equal to the total momentum after the event.

2.4.2 Impulse

Definition: Impulse is the product of force and the time over which it acts, $J = F.\Delta t$

Impulse-Momentum Theorem: Impulse causes a change in momentum, expressed as $J = \Delta p$.

2.5 Gravity

Gravity is a fundamental force that attracts objects toward one another. It plays a crucial role in the motion of planets, stars, and galaxies, as well as objects on Earth.

2.5.1 Universal Law of Gravitation

According to Newton's law of gravitation, any two masses exert an attractive force on each other. The formula is:

$$F = G\frac{m_1.m_2}{r^2}$$

F = gravitational force

G = gravitational constant

m_1 and m_2 = masses of the objects

r = distance between the centers of the masses

2.5.2 Weight and Gravitational Acceleration

Weight is the force of gravity acting on an object's mass:

$$W = mg$$

W = weight

m = mass

g = acceleration due to gravity, approximately 9.8, m/s² on Earth

2.6 Waves and Oscillations

Waves are disturbances that transfer energy from one place to another, and oscillations refer to back-and-forth motion around an equilibrium position.

2.6.1 Types of Waves

Mechanical Waves: Require a medium (e.g., sound waves in air).

Electromagnetic Waves: Do not require a medium and can travel through a vacuum (e.g., light waves).

2.6.2 Properties of Waves

Wavelength (λ): Distance between two successive points in phase (e.g., crest to crest).

Frequency (f): Number of wave cycles per second, measured in hertz (Hz).

Amplitude: Height of the wave from its equilibrium position.

Speed: The speed of a wave is calculated by $v = f.\lambda$

2.7 Electric and Magnetic Fields

Electric and magnetic fields are forces that affect charged particles and play a central role in electromagnetism.

2.7.1 Electric Fields

Definition: A region around a charged particle where an electric force is exerted on other charges.

Field Strength: Defined as $E = \frac{F}{q}$, where E is the electric field strength, F is force, and q is charge.

2.7.2 Magnetic Fields

Definition: A region around a magnetic material or a moving electric charge where magnetic force is exerted.

Magnetic Force: Moving charges produce magnetic fields, which can exert forces on other moving charges or magnetic materials.

2.8 Summary

In this chapter, we covered the core concepts in physics that form the basis for understanding more advanced topics. From motion and forces to energy, momentum, gravity, waves, and fields, these principles explain the fundamental behaviors of the physical world. With this understanding, you're ready to explore these ideas in greater detail in the chapters that follow, building a deeper understanding of how the universe operates.

Chapter 3: Tools and Resources for Physics

To effectively study and experiment with physics, it's essential to have the right tools and resources. This chapter provides an overview of the fundamental tools physicists and students use, from basic lab equipment to advanced resources, digital tools, and online resources that make physics accessible and engaging.

3.1 Laboratory Equipment and Measurement Tools

Physics relies heavily on precise measurements, so having the right equipment is crucial for conducting experiments accurately. Here's a breakdown of some essential laboratory tools:

3.1.1 Basic Measurement Instruments

Ruler and Measuring Tape: For measuring length and distance, often in centimeters or meters. Essential for experiments involving displacement, height, and depth.

Calipers and Micrometers: Precision instruments for measuring small distances with high accuracy, often used to measure the diameter of objects or the thickness of materials.

Protractor: For measuring angles, especially in mechanics and optics experiments.

Stopwatch: Used to measure time intervals, an essential tool in experiments involving speed, velocity, and acceleration.

Balance: For measuring mass, balances are crucial for experiments involving forces, weight, and density calculations.

3.1.2 Force and Motion Measurement

Spring Scale: Measures force in newton's by assessing the stretch of a spring when weight is applied.

Accelerometer: A device that measures acceleration, often used in experiments to determine forces on moving objects.

Ticker Timer: A device that marks intervals on a moving tape, allowing for analysis of speed, acceleration, and time in mechanics experiments.

3.1.3 Electrical Measurement Instruments

Multimeter: Measures voltage, current, and resistance, fundamental for experiments in electromagnetism and circuits.

Oscilloscope: Visualizes electrical signals, showing how voltage changes over time. Essential for studying waveforms and AC circuits.

Ammeter and Voltmeter: Specifically measure electric current (amperes) and voltage (volts), respectively, in a circuit.

Power Supply: Provides a stable voltage or current source for electrical experiments and circuit testing.

3.1.4 Optics and Light Measurement

Optical Bench: Used for experiments involving lenses, mirrors, and light sources to measure focal lengths and study image formation.

Spectrometer: Measures properties of light, including wavelength and intensity, essential for experiments in optics and quantum mechanics.

Photometer: Measures the intensity of light, useful in experiments studying brightness, energy, and light absorption.

3.1.5 Temperature Measurement

Thermometer: Measures temperature in degrees Celsius or Fahrenheit, useful in thermodynamics.

Thermocouple: A device that measures temperature using two wires of different metals, often used in industrial and laboratory settings for precise temperature readings.

Infrared Thermometer: Measures temperature from a distance using infrared radiation, useful for measuring surface temperatures without contact.

3.2 Mathematics in Physics

Mathematics is the language of physics. Several mathematical tools are essential for solving physics problems:

3.2.1 Algebra and Trigonometry

Used to solve equations, rearrange formulas, and analyze relationships. Algebra is essential for handling equations of motion, while trigonometry is crucial for understanding angles in force vectors, waves, and optics.

3.2.2 Calculus

Differentiation: Used to determine rates of change, such as velocity from displacement or acceleration from velocity.

Integration: Used to calculate areas under curves, such as work done by a variable force or distance covered when velocity varies over time.

Calculus is especially useful in advanced mechanics, electromagnetism, and thermodynamics.

3.2.3 Vectors and Geometry

Physics often deals with quantities that have both magnitude and direction, such as force, velocity, and acceleration. Vector algebra helps us add, subtract, and resolve forces into components.

Geometry is fundamental in analyzing shapes, distances, and angles in optics and electromagnetism.

3.2.4 Graphs and Data Analysis

Graphs visually represent relationships between variables, making them useful for analyzing trends, such as how velocity changes over time or how force relates to acceleration.

Data Analysis: Tools like standard deviation and error analysis help determine the reliability of results and the precision of measurements.

3.3 Software Tools and Simulation Resources

Digital tools and simulations provide a convenient way to study physics concepts, especially for complex systems that are difficult to observe directly.

3.3.1 Data Collection and Analysis Software

Excel or Google Sheets: Useful for recording, analyzing, and plotting data. You can calculate averages, standard deviations, and even simulate simple models using formulas.

MATLAB and Mathematica: Advanced mathematical software for modeling, visualizing, and analyzing complex data. Useful for higher-level physics studies in areas like quantum mechanics and electromagnetic fields.

Python: A programming language with libraries like NumPy and SciPy, which are powerful tools for data analysis, scientific computing, and creating simulations.

3.3.2 Simulation Software

PhET Interactive Simulations: Free, interactive physics simulations for students to visualize concepts like motion, waves, circuits, and quantum phenomena.

Algodoo: A 2D simulation tool where you can create, test, and observe physical scenarios, especially useful for understanding mechanics.

VPython: A Python module that allows users to create 3D simulations of physical systems, ideal for visualizing motion and electromagnetic fields.

3.3.3 Circuit Design Software

CircuitLab and Tinkercad Circuits: Web-based platforms for designing and simulating electrical circuits. Great for experimenting with circuits before building them physically.

Multisim: Advanced circuit simulation software widely used in academic and engineering settings to model and analyze electronic circuits.

3.4 Online Resources for Physics Learning

The internet offers a wealth of resources for learning physics, from online courses to video tutorials. Here are some top resources:

3.4.1 Educational Platforms

Khan Academy: Free online courses covering a broad range of physics topics with video lectures, practice problems, and quizzes.

Coursera and edX: These platforms offer university-level courses on physics topics, often taught by professors from institutions like MIT, Stanford, and Harvard. Some courses are free or low-cost.

MIT OpenCourseWare: Provides free access to materials from MIT's physics courses, including lecture notes, assignments, and exams.

3.4.2 Video Tutorials and Channels

YouTube Channels:

- Physics Girl: Fun, engaging videos on physics concepts and experiments.
- MinutePhysics: Short, animated videos explaining complex physics topics in an accessible way.
- Veritasium: Covers a wide range of science topics, including physics, with a focus on real-world applications and thought-provoking questions.

CrashCourse Physics: A structured video series that covers high school and introductory college-level physics in an engaging, easy-to-follow format.

3.4.3 Interactive and Virtual Labs

Labster: Virtual labs that provide a hands-on experience in physics, allowing you to conduct experiments in a safe, simulated environment.

OLabs: Interactive online labs aligned with school-level physics curricula, ideal for students who want to perform experiments at home.

3.5 Reference Books and Textbooks

While online resources are valuable, classic textbooks provide structured, in-depth explanations of physics concepts. Here are some recommended physics texts:

"Fundamentals of Physics" by Halliday, Resnick, and Walker: A widely used textbook covering all the basics of physics, with practice problems and detailed explanations.

"Conceptual Physics" by Paul G. Hewitt: Great for beginners, this book emphasizes understanding physics concepts without heavy math.

"The Feynman Lectures on Physics" by Richard Feynman: An engaging series of lectures that cover a broad range of physics topics, often used as a supplementary text.

"University Physics" by Young and Freedman: A comprehensive textbook that covers physics topics in detail, often used in university courses.

3.6 Apps and Mobile Tools

Mobile apps can make learning physics more interactive and accessible. Here are some recommended apps:

Physics Toolbox Suite: Uses your smartphone's sensors to measure data like acceleration, magnetic field strength, sound levels, and more. Great for hands-on experiments.

Pocket Physics: A quick-reference app covering fundamental concepts, equations, and diagrams across all physics topics.

Phyphox: An app that utilizes the sensors in your smartphone for real-time data collection and analysis. It's useful for experiments involving acceleration, magnetism, and sound.

SimPHY: A physics simulation app that lets you model motion, forces, electricity, and magnetism with 2D and 3D simulations.

3.7 Conclusion

This chapter introduced the essential tools and resources for studying physics, from fundamental lab equipment to advanced simulation software, online courses, and reference materials. Armed with these tools, you'll be able to explore physics more deeply, apply concepts to real-world scenarios, and conduct meaningful experiments.

Chapter 4: Exploring Mechanics

Mechanics is one of the oldest and most fundamental branches of physics, concerned with the motion of objects and the forces that cause these motions. It provides the foundation for understanding how things move and interact in the physical world, from everyday objects to celestial bodies. In this chapter, we will explore the essential concepts in mechanics, including motion, forces, energy, and momentum. We will also examine how these concepts are applied in real-world situations and experiments.

4.1 What is Mechanics?

Mechanics is the study of how and why objects move. It is generally divided into two main areas:

Kinematics: The study of motion itself, without considering the forces that cause it. Kinematics focuses on describing how objects move, their position, velocity, and acceleration.

Dynamics: The study of the forces that cause motion. Dynamics explores how forces influence the motion of objects and how they change their velocity or direction.

Together, these two areas help us understand everything from a car moving on a road to the planets orbiting the Sun.

4.2 Types of Motion

Understanding the types of motion is crucial to studying mechanics. Different objects may experience different kinds of motion, and each type has specific characteristics.

4.2.1 Linear Motion

Linear motion occurs when an object moves along a straight path. This is the simplest type of motion and is often described by variables such as velocity, displacement, and acceleration.

Example: A car moving down a straight road.

4.2.2 Rotational Motion

Rotational motion refers to the motion of objects that rotate around an axis. A classic example is the Earth's rotation around its axis. Important quantities include angular displacement, angular velocity, and angular acceleration.

Example: A spinning wheel or the Earth's daily rotation.

4.2.3 Projectile Motion

Projectile motion describes the motion of an object that is thrown or projected into the air, influenced only by gravity and air resistance (though air resistance is often neglected for simplicity).

Example: A ball thrown into the air.

Key components: Horizontal motion (constant velocity) and vertical motion (accelerated due to gravity).

4.3 Energy and Work in Mechanics

Energy is the ability to do work, and work itself is the transfer of energy. In mechanics, energy and work are closely related to the motion of objects.

4.3.1 Work

Work is done when a force causes an object to move. The amount of work is calculated as:

- $W = F \cdot d \cdot \cos(\theta)$
  - W = Work
  - F = force applied
  - d = displacement
  - θ = angle between force and displacement direction

If the force and displacement are in the same direction, the work done is positive. If they are in opposite directions, the work done is negative.

4.3.2 Kinetic Energy

Kinetic energy is the energy of an object due to its motion and is given by the equation:

- $KE = \frac{1}{2}mv^2$
  - m = mass of object
  - v = velocity of object

Kinetic energy increases with both mass and the square of velocity. This is why faster-moving objects have much more kinetic energy than slower ones.

4.3.3 Potential Energy

Potential energy is stored energy that an object possesses due to its position or condition. The most common form of potential energy is gravitational potential energy, given by:

- $PE = mgh$
  - m = mass of the object
  - g = acceleration due to gravity
  - h = height of the object above the ground

When you lift an object, you are doing work to increase its potential energy. The higher the object, the more gravitational potential energy it has.

4.3.4 Conservation of Energy

The law of conservation of energy states that energy cannot be created or destroyed, only converted from one form to another. For example, a roller coaster converts potential energy into kinetic energy as it drops down a track. At any point, the total energy (kinetic + potential) remains constant, provided no external forces like friction are acting on the system.

4.4 Rotational Motion and Torque

While linear motion is the most common type of motion we experience, many objects in the real world rotate. Understanding rotational motion is important for applications like engines, wheels, and even planetary motion.

4.4.1 Torque

Torque is the rotational equivalent of force. It is the force that causes an object to rotate about an axis. The formula for torque is:

- $\tau = r \cdot F \cdot \sin\theta$
  - τ = torque
  - r = distance from the axis of rotation
  - F = force applied
  - θ = angle between the force and the line connecting the point of application to the axis of rotation

Torque depends not only on the magnitude of the force but also on where the force is applied relative to the axis of rotation.

4.4.2 Moment of Inertia

Moment of inertia is the rotational equivalent of mass and is a measure of an object's resistance to rotational acceleration. It depends on both the mass of the object and how that mass is distributed relative to the axis of rotation.

4.4.3 Angular Velocity and Acceleration

Angular velocity is the rate of change of the angle of rotation, and angular acceleration is the rate of change of angular velocity. These quantities are similar to linear velocity and acceleration in translational motion.

- $\omega = \frac{\Delta\theta}{\Delta t}$ (angular velocity)
- $a = \frac{\Delta\omega}{\Delta t}$ (angular acceleration)

4.5 Applications of Mechanics

Mechanics has a wide range of applications, from the everyday movement of objects to complex engineering systems.

Engineering: Mechanics is essential in the design and construction of structures, machinery, vehicles, and airplanes. The principles of force, motion, and energy are used to ensure safety, efficiency, and performance.

Sports: In sports like baseball or soccer, mechanics is used to analyze the forces and motions involved in hitting, kicking, and throwing.

Astronomy: The motion of celestial bodies, such as planets, moons, and comets, can be described using the principles of mechanics.

4.6 Summary

In this chapter, we explored the fundamental principles of mechanics, including motion, forces, energy, and momentum. Understanding these core concepts is essential for explaining the behavior of objects in motion, from everyday experiences to advanced scientific applications. With this foundational knowledge, you are prepared to delve deeper into the more complex aspects of physics, where mechanics plays a critical role in explaining how the universe works.

This chapter has provided a comprehensive introduction to the key topics in mechanics. By mastering these principles, you will be able to understand the physical world and apply this knowledge to practical problems in engineering, physics, and beyond.

Chapter 5: Understanding Thermodynamics

Thermodynamics is the branch of physics that deals with the study of heat, work, temperature, and energy. It explores how energy is transferred within systems and how these transformations affect matter. Thermodynamics plays a crucial role in a range of fields, from physics and chemistry to engineering and environmental science. This chapter delves into the foundational principles of thermodynamics, the laws that govern it, and its real-world applications.

5.1 What is Thermodynamics?

Thermodynamics is the science of energy transformations, focusing primarily on how thermal energy (heat) is converted to other forms of energy, such as work. It studies the behavior of macroscopic systems based on their thermal properties. Thermodynamics helps answer questions like:

- How does heat transfer between objects?
- How much work can be extracted from a heat engine?
- What is the maximum efficiency of a system?

At its core, thermodynamics is governed by several key concepts and laws that define the relationship between heat, work, and energy.

5.2 Key Concepts in Thermodynamics

To understand thermodynamics, it's essential to familiarize yourself with its key concepts and the quantities involved in analyzing energy transformations.

5.2.1 System and Surroundings

System: The part of the universe being studied, which could be as simple as a cup of coffee or as complex as an engine.

Surroundings: Everything outside the system that interacts with it. The boundary separates the system from the surroundings and can be open, closed, or isolated:

- Open System: Can exchange both matter and energy with the surroundings.
- Closed System: Exchanges only energy, not matter, with the surroundings.
- Isolated System: Neither energy nor matter is exchanged with the surroundings.

5.2.2 Heat and Temperature

Heat (Q): A form of energy transfer that occurs between systems due to a temperature difference. Heat flows naturally from warmer objects to cooler ones.

Temperature (T): A measure of the average kinetic energy of particles in a substance. It determines the direction of heat transfer.

5.2.3 Work (W)

Work is the energy transferred to or from a system by external forces. In thermodynamics, work is often associated with the expansion or compression of gases.

5.2.4 Internal Energy (U)

The total energy of a system, which includes the kinetic and potential energies of all particles within the system. Internal energy can change as a result of heat transfer or work done on/by the system.

5.3 The Laws of Thermodynamics

Thermodynamics is governed by four fundamental laws that describe how energy is conserved, transferred, and transformed. These laws form the foundation of all thermodynamic processes.

5.3.1 The Zeroth Law of Thermodynamics

Statement: If two systems are each in thermal equilibrium with a third system, then they are in thermal equilibrium with each other.

Implication: This law establishes the concept of temperature and allows us to measure it. It essentially states that temperature is a property that can define the state of thermal equilibrium.

5.3.2 The First Law of Thermodynamics (Conservation of Energy)

Statement: The change in internal energy of a system (ΔU) is equal to the heat added to the system (Q) minus the work done by the system (W).

Mathematical Formulation:

$$\Delta U = Q - W$$

Implication: The first law is essentially a restatement of the conservation of energy. It tells us that energy cannot be created or destroyed, only transferred or transformed.

Example: When a gas is compressed in a cylinder, the work done on the gas increases its internal energy, which may result in a rise in temperature.

5.3.3 The Second Law of Thermodynamics (Entropy)

Statement: In any natural thermodynamic process, the total entropy of a system and its surroundings tends to increase over time. Entropy is a measure of disorder or randomness.

Implication: The second law introduces the concept of irreversibility. Heat flows spontaneously from hot objects to cold objects, and without external work, processes tend toward greater disorder (higher entropy).

Example: When you place an ice cube in a warm drink, heat flows from the liquid to the ice, melting it and increasing the system's entropy.

5.3.4 The Third Law of Thermodynamics (Absolute Zero)

Statement: As the temperature of a system approaches absolute zero, the entropy of the system approaches a minimum, constant value.

Implication: This law suggests that it is impossible to reach absolute zero, as doing so would require removing all thermal energy from the system.

Example: Absolute zero (0 Kelvin or -273.15°C) is a theoretical point where atomic motion nearly ceases. Although temperatures close to absolute zero have been achieved in labs, absolute zero itself is unreachable.

5.4 Thermodynamic Processes

Thermodynamic processes describe how a system changes from one state to another. These processes are essential for understanding engines, refrigerators, and other systems where energy transfer is involved.

5.4.1 Isothermal Process

Definition: A process that occurs at constant temperature.

Example: When an ideal gas expands slowly enough that its temperature remains constant, the internal energy stays the same, and any heat added to the system is converted into work.

5.4.2 Adiabatic Process

Definition: A process where no heat is transferred into or out of the system.

Example: When a gas is compressed quickly, there is no time for heat exchange, so all work done on the gas increases its internal energy, raising its temperature.

5.4.3 Isobaric Process

Definition: A process that occurs at constant pressure.

Example: When a gas is heated at constant pressure, it expands, doing work on the surroundings.

5.4.4 Isochoric Process (Isovolumetric)

Definition: A process where volume remains constant, meaning no work is done.

Example: Heating a gas in a sealed, rigid container increases its pressure but does not change its volume.

5.5 Understanding Entropy and the Arrow of Time

Entropy is a measure of the disorder or randomness in a system. In thermodynamic processes, entropy tends to increase, a principle that provides an "arrow of time," suggesting that time flows in one direction, from order to disorder.

Reversible Process: A hypothetical process that can occur in both directions without any change in entropy. In reality, perfect reversibility is impossible because it would require no friction or other dissipative effects.

Irreversible Process: All natural processes are irreversible and lead to an increase in entropy. For example, mixing two gases is irreversible; once mixed, they cannot separate naturally.

Entropy provides insight into why certain processes, like heat spontaneously flowing from hot to cold, occur naturally, while the reverse requires external intervention.

5.6 Applications of Thermodynamics

Thermodynamics has practical applications across various fields, from engineering and chemistry to biology and environmental science.

5.6.1 Heat Engines

A heat engine converts thermal energy into mechanical work. It operates by absorbing heat from a high-temperature source, doing work, and releasing waste heat to a low-temperature sink. Examples include car engines and steam turbines.

Efficiency of a Heat Engine: The efficiency of a heat engine is the ratio of the work it does to the heat absorbed.

$$\eta = \frac{W}{Q_{in}}$$

Carnot efficiency, which represents the maximum efficiency, is determined by the temperatures of the hot and cold reservoirs.

5.6.2 Refrigerators and Heat Pumps

Refrigerators and heat pumps use work to transfer heat from a cooler area to a warmer one. A refrigerator removes heat from its interior and releases it outside, keeping the inside cold.

Heat pumps, in contrast, can warm or cool a space by reversing the direction of heat flow, which is useful for both heating and air conditioning systems.

5.6.3 Biological Systems

Thermodynamics plays a critical role in biology. Metabolism in living organisms can be thought of as a thermodynamic process where chemical energy in food is converted into energy for cellular work.

Enzymes act as catalysts to speed up biochemical reactions, helping organisms maintain low entropy levels despite the tendency toward disorder.

5.6.4 Environmental Science and Climate

The Earth's climate system can be analyzed thermodynamically by studying how energy from the sun is absorbed, retained, and re-radiated. Understanding thermodynamics is crucial for analyzing heat exchanges in the atmosphere and predicting climate change.

5.7 Summary

Thermodynamics is a powerful framework for understanding how energy is transformed and transferred. Through its laws and concepts, thermodynamics explains why energy flows in certain directions, why entropy increases, and how we can harness energy to perform work. Whether in heat engines, biological systems, or environmental processes, the principles of thermodynamics shape much of the world around us.

Chapter 6: Fundamentals of Electromagnetism

Electromagnetism is the branch of physics that studies the interactions between electric charges and magnetic fields. This field unites electricity and magnetism into a single theory of the electromagnetic force, one of the four fundamental forces of nature. Electromagnetism is responsible for many phenomena we observe daily, from the behavior of magnets to the operation of electrical devices and wireless communication. This chapter will cover the fundamental principles of electromagnetism, including electric fields, magnetic fields, electromagnetic waves, and practical applications.

6.1 What is Electromagnetism?

Electromagnetism describes the interaction between charged particles through electric and magnetic fields. A key concept of electromagnetism is that a changing electric field can create a magnetic field, and a changing magnetic field can create an electric field. This relationship, known as electromagnetic induction, is the foundation of numerous technologies, from electric motors to transformers.

In general, electromagnetism is governed by Maxwell's equations, a set of four mathematical equations that describe how electric and magnetic fields are generated and altered by charges and currents.

6.2 Key Concepts in Electromagnetism

To understand electromagnetism, we need to start with the basic building blocks: electric charges, electric fields, magnetic fields, and how they interact with one another.

6.2.1 Electric Charge

Definition: Electric charge is a fundamental property of particles that causes them to experience a force when placed in an electric or magnetic field. There are two types of electric charge: positive and negative.

Units: The unit of charge is the coulomb (C).

Properties: Opposite charges attract, and like charges repel. This principle is the basis of electrostatic force.

6.2.2 Electric Fields (E-Fields)

Definition: An electric field is a region around a charged object where other charges experience a force. Electric fields are represented by field lines, which indicate the direction a positive test charge would move within the field.

Electric Field Strength (E): The strength of an electric field at any point is given by $E = \frac{F}{q}$, where F is the force experienced by a test charge q.

Coulomb's Law: The force between two charges q_1 and q_2 separated by a distance r is given by: $F = k \frac{q_1 q_2}{r^2}$

6.2.3 Magnetic Fields (B-Fields)

Definition: A magnetic field is a region where a magnetic force can be observed, typically generated by moving electric charges (currents) or magnetic materials.

Magnetic Field Strength (B): Magnetic field strength is measured in teslas (T).

Magnetic Field Lines: Magnetic fields are represented by lines that flow from the north pole to the south pole of a magnet, indicating the field's direction.

6.2.4 Electromagnetic Force

The electromagnetic force encompasses both electric and magnetic forces. A charged particle in an electric field experiences an electrostatic force, while a moving charge in a magnetic field experiences a magnetic force. The combined effects of these forces are fundamental to understanding electromagnetic interactions.

6.3 Maxwell's Equations

Maxwell's Equations are a set of four equations that provide a complete description of the behavior of electric and magnetic fields. They are essential to understanding how electromagnetic waves propagate.

1. **Gauss's Law:** The electric flux through any closed surface is proportional to the total electric charge within the surface.

$$\nabla \cdot E = \frac{\rho}{\epsilon_0}$$

2. **Gauss's Law for Magnetism:** The magnetic flux through any closed surface is zero, indicating that magnetic monopoles do not exist.

$$\nabla \cdot B = 0$$

3. **Faraday's Law of Induction:** A changing magnetic field creates an electric field. This law is the basis for electromagnetic induction, which allows generators to convert mechanical energy into electrical energy.

$$\nabla \times E = -\frac{\partial B}{\partial t}$$

4. **Ampère's Law (with Maxwell's addition):** A changing electric field or electric current generates a magnetic field.

$$\nabla \times B = \mu_0 J + \mu_0 \epsilon_0 \frac{\partial E}{\partial t}$$

Together, these equations describe how electric and magnetic fields interact and propagate through space.

6.4 Electromagnetic Induction

Electromagnetic induction is the process by which a changing magnetic field creates an electric current. This principle, discovered by Michael Faraday, is fundamental to the operation of generators, transformers, and other electrical devices.

Faraday's Law of Induction: The induced electromotive force (emf) in a closed loop is proportional to the rate of change of magnetic flux through the loop.

$$emf = -\frac{d\Phi_B}{dt}$$

Lenz's Law: The direction of the induced current will oppose the change in magnetic flux that produced it. This is a consequence of the conservation of energy and ensures that the induced current resists the change in the magnetic field.

Applications: Electromagnetic induction is utilized in power generation, where mechanical energy (like rotating a turbine) is used to create electricity. It also powers induction cooking and wireless charging technologies.

6.5 Electromagnetic Waves

Electromagnetic waves are oscillations of electric and magnetic fields that propagate through space. Unlike sound or water waves, electromagnetic waves do not require a medium and can travel through a vacuum at the speed of light (approximately).

Properties of Electromagnetic Waves:

Electromagnetic waves are transverse, meaning that the electric and magnetic fields oscillate perpendicular to the direction of wave propagation.

The frequency and wavelength of an electromagnetic wave determine its energy and where it lies on the electromagnetic spectrum.

Electromagnetic Spectrum:

The electromagnetic spectrum ranges from low-frequency radio waves to high-frequency gamma rays. It includes, in order of increasing frequency: radio waves, microwaves, infrared radiation, visible light, ultraviolet radiation, X-rays, and gamma rays.

Different parts of the spectrum have unique properties and applications. For example, visible light allows us to see, while X-rays are used in medical imaging.

6.6 Practical Applications of Electromagnetism

Electromagnetism has countless applications in technology, medicine, and everyday life. Here are a few notable applications:

6.6.1 Electric Motors

Electric motors use electromagnetic induction to convert electrical energy into mechanical energy. In a motor, an electric current flows through a coil within a magnetic field, producing a force that rotates the coil. This rotation can then be harnessed to perform work, powering everything from household appliances to industrial machinery.

6.6.2 Generators

Generators work on the principle of electromagnetic induction, converting mechanical energy into electrical energy. In a generator, a coil is rotated within a magnetic field, inducing a current in the coil. This is the fundamental mechanism behind power generation in hydroelectric, wind, and thermal power plants.

6.6.3 Transformers

Transformers are devices that increase or decrease the voltage of alternating current (AC) in a circuit. They operate on the principle of electromagnetic induction, with a primary coil and a secondary coil wound around a magnetic core. Transformers are crucial for efficiently transmitting electricity over long distances by stepping up the voltage for transmission and stepping it down for safe usage.

6.6.4 Wireless Communication

Electromagnetic waves enable wireless communication, including radio, television, and cellular networks. In wireless communication, information is encoded in electromagnetic waves, which are transmitted over distances. The receiving device decodes these waves back into usable information, such as audio, video, or data.

6.6.5 Medical Imaging

Electromagnetic waves are vital in medical imaging technologies. X-rays, for instance, penetrate soft tissues but are absorbed by denser materials like bones, creating images for diagnostic purposes. MRI (Magnetic Resonance Imaging) uses strong magnetic fields and radio waves to create detailed images of internal body structures.

6.7 Summary

Electromagnetism is one of the pillars of modern physics and technology. By understanding the behavior of electric and magnetic fields, we gain insights into the forces that govern the universe at both microscopic and macroscopic scales. The principles of electromagnetism enable power generation and transmission, wireless communication, electric motors, medical imaging (like MRI), data storage, wireless charging, and the operation of electronics. They underpin most modern technologies by allowing us to control and utilize electric and magnetic fields for diverse applications in energy, computing, and healthcare.

Chapter 7: Waves and Optics Essentials

Waves and optics form a fundamental part of physics, dealing with the behavior, properties, and applications of waves and light. This chapter explores the essential concepts of wave mechanics and optics, covering the properties of waves, the nature of light, and phenomena such as reflection, refraction, diffraction, and interference. These principles are essential for understanding how sound travels, how we see, and how technologies like lasers, lenses, and fiber optics work.

7.1 What is a Wave?

A wave is a disturbance that travels through a medium or space, transferring energy from one point to another without transferring matter. Waves are found everywhere, from sound and water waves to electromagnetic waves, including light. Waves are often characterized by their frequency, wavelength, amplitude, and speed.

Types of Waves

1. **Mechanical Waves:** These require a medium (such as water, air, or solids) to travel through. Examples include sound waves and water waves.

2. **Electromagnetic Waves:** These do not require a medium and can travel through a vacuum. Light, radio waves, and X-rays are electromagnetic waves.

3. **Transverse Waves:** In these waves, particles of the medium move perpendicular to the direction of wave propagation, like in light waves or water waves.

4. **Longitudinal Waves:** In these waves, particles of the medium move parallel to the direction of wave propagation. Sound waves are longitudinal.

7.2 Properties of Waves

To understand how waves work, it's crucial to be familiar with their basic properties:

7.2.1 Wavelength (λ)

The distance between two consecutive crests or troughs in a wave.

Wavelength is measured in meters (m) and is inversely proportional to frequency.

7.2.2 Frequency (f)

The number of wave cycles that pass a given point per second. Frequency is measured in hertz (Hz).

Higher frequency means a shorter wavelength and vice versa.

7.2.3 Amplitude (A)

The maximum displacement of points on a wave from its rest position. Amplitude determines the wave's energy and intensity.

Larger amplitude waves carry more energy.

7.2.4 Wave Speed (v)

The speed at which a wave propagates through a medium.

The wave speed can be calculated as:

$v = f\lambda$

7.2.5 Phase

Phase describes the position of a point in the wave cycle. Waves in phase reinforce each other, while waves out of phase can cancel each other out (interference).

7.3 Sound Waves

Sound waves are longitudinal mechanical waves that require a medium to travel through. They are produced by vibrations and are typically perceived as compressions and rarefactions in the air.

Characteristics of Sound Waves

Pitch: Determined by frequency; higher frequencies result in higher pitches.

Volume: Determined by amplitude; greater amplitude means louder sounds.

Speed of Sound: Varies depending on the medium (faster in solids, slower in gases).

Sound waves undergo phenomena such as reflection, refraction, and interference, allowing us to hear and perceive direction.

7.4 Light as an Electromagnetic Wave

Light is an electromagnetic wave, which means it does not require a medium and can travel through a vacuum. Light exhibits both wave and particle characteristics, a concept known as wave-particle duality. It travels at approximately in a vacuum.

The Nature of Light

Visible Light: The portion of the electromagnetic spectrum visible to the human eye, ranging from 400 nm (violet) to 700 nm (red).

Wave-Particle Duality: Light behaves as both a wave (showing interference and diffraction) and a particle (photon) in different circumstances.

7.5 Basic Concepts of Optics

Optics is the branch of physics that studies the behavior of light and its interactions with matter. It focuses on phenomena like reflection, refraction, diffraction, and interference.

7.5.1 Reflection

Reflection occurs when light bounces off a surface. The law of reflection states that:

- The angle of incidence (incoming angle) is equal to the angle of reflection (outgoing angle).
  - Mirrors, for example, work by reflecting light to form images.

Types of Reflection

Specular Reflection: Occurs on smooth surfaces like mirrors, where parallel rays are reflected in a single direction.

Diffuse Reflection: Occurs on rough surfaces, where light is scattered in many directions.

7.5.2 Refraction

Refraction is the bending of light as it passes from one medium to another, due to a change in speed. The degree of bending is governed by the refractive indices of the media and is described by Snell's Law:

$$n_1 \sin\theta_1 = n_2 \sin\theta_2$$

Applications of Refraction:

Lenses: Used in eyeglasses, microscopes, and cameras to focus light and create clear images.

Prisms: Split white light into its component colors through refraction.

7.5.3 Diffraction

Diffraction is the bending and spreading of waves around obstacles or through narrow openings. The extent of diffraction depends on the wavelength and the size of the obstacle or aperture.

Example: Sound waves diffract more around corners than light waves because of their longer wavelengths.

Applications: Diffraction grating is used spectrometers to analyze the composition of light by spreading it into a spectrum.

7.5.4 Interference

Interference occurs when two waves meet and combine to form a new wave pattern. It can be constructive or destructive:

Constructive Interference: When waves are in phase and their amplitudes add up, creating a larger wave.

Destructive Interference: When waves are out of phase, canceling each other out.

Applications:

Thin Film Interference: Seen in soap bubbles and oil films, where light reflects from multiple layers, creating colorful patterns.

Interferometry: Used in precision measurements, such as detecting gravitational waves.

7.6 Optical Instruments

Optical instruments use lenses, mirrors, and other components to manipulate light. Some commonly used optical instruments include:

7.6.1 Microscopes

Microscopes use multiple lenses to magnify small objects, allowing us to see details not visible to the naked eye. They are crucial tools in biology, materials science, and medical research.

7.6.2 Telescopes

Telescopes gather and focus light to magnify distant objects, such as stars and galaxies. They are essential tools in astronomy and come in different types, including refracting telescopes (using lenses) and reflecting telescopes (using mirrors).

7.6.3 Cameras

Cameras capture light through a lens, focusing it on a sensor or film to produce an image. Modern cameras use multiple lenses to control zoom, focus, and image clarity, making them versatile tools in photography and film.

7.6.4 Lasers

Lasers emit light through stimulated emission, producing a coherent, focused beam. Lasers have applications in medicine (surgery), communication (fiber optics), and entertainment (laser shows).

7.7 The Electromagnetic Spectrum

The electromagnetic spectrum covers a range of frequencies and wavelengths, from low-energy radio waves to high-energy gamma rays. Each type of wave has unique properties and applications:

Radio Waves: Used in broadcasting, communications, and radar.

Microwaves: Used in cooking and satellite communication.

Infrared Radiation: Emitted by warm objects, used in night vision and remote controls.

Visible Light: The only part of the spectrum visible to humans, used in everything from lighting to vision.

Ultraviolet Light: Used in sterilization and fluorescent lights.

X-rays: Used in medical imaging and security scanning.

Gamma Rays: Produced by radioactive atoms, used in cancer treatment and nuclear research.

7.8 Practical Applications of Waves and Optics

Waves and optics are essential in many technologies and fields:

1. **Medical Imaging:** Ultrasound uses sound waves for imaging, while X-rays and MRI utilize electromagnetic waves.

2. **Communication:** Radio, television, and cellular communication rely on radio waves, while fiber optics use light waves for high-speed data transmission.

3. **Entertainment:** Sound and light waves are fundamental in audio-visual technologies like speakers, projectors, and theaters.

4. **Environmental Sensing:** Optical instruments and wave-based sensors are used in weather prediction, environmental monitoring, and geological surveying.

7.9 Summary

Waves and optics provide the foundation for understanding the behavior and interaction of different forms of energy. By exploring the properties of waves, the nature of light, and how these interact with various materials, we gain insights into everything from the mechanics of sound to the behavior of light and its interactions with matter.

Chapter 8: Introduction to Modern Physics

Modern physics is a branch of physics that emerged in the early 20th century, fundamentally changing our understanding of nature. Unlike classical physics, which deals with macroscopic objects and familiar phenomena, modern physics delves into the strange and often counterintuitive world of the very small and the very fast. This chapter introduces key concepts of modern physics, including quantum mechanics, relativity, atomic structure, and fundamental particles. Understanding these principles provides insights into the nature of reality itself and underpins much of today's technological advancements, from electronics to nuclear energy.

8.1 What is Modern Physics?

Modern physics arose from discoveries that classical mechanics and electromagnetism could not explain, such as the behavior of atoms and the speed of light. The two primary pillars of modern physics are:

Quantum Mechanics: The study of particles on an atomic and subatomic scale, where phenomena like uncertainty, superposition, and wave-particle duality are central.

Theory of Relativity: Developed by Albert Einstein, relativity redefined our concepts of space, time, and gravity.

Together, these theories reshape how we understand the universe on a fundamental level.

8.2 Key Concepts in Modern Physics

8.2.1 Quantum Mechanics

Quantum mechanics studies matter and energy at the smallest scales, where the laws of classical physics no longer apply. Key concepts in quantum mechanics include:

Quantization: Energy exists in discrete units, called quanta, rather than being continuous. For example, photons are quanta of light energy.

Wave-Particle Duality: Particles like electrons exhibit properties of both particles and waves, depending on how they are observed. This duality is fundamental in explaining phenomena such as interference patterns.

Uncertainty Principle: Proposed by Werner Heisenberg, this principle states that certain pairs of physical properties, like position and momentum, cannot be precisely measured at the same time. The more accurately we know one property, the less accurately we can know the other.

Superposition: Particles exist in all possible states simultaneously until measured. This principle is famously illustrated by Schrödinger's cat thought experiment, where a cat in a closed box is considered both alive and dead until observed.

Applications of Quantum Mechanics: Quantum mechanics is the foundation of technologies such as semiconductors, lasers, MRI machines, and quantum computing.

8.2.2 Theory of Relativity

Relativity, developed by Einstein, consists of two main parts:

Special Relativity: Focuses on objects moving at constant high speeds, near the speed of light. Key results include time dilation, length contraction, and the famous equation , which shows that mass and energy are interchangeable.

General Relativity: A theory of gravity that describes gravity as the curvature of space-time caused by massive objects. General relativity explains phenomena such as gravitational waves, black holes, and the bending of light by gravity.

Applications of Relativity: Relativity is essential in fields such as GPS technology, astrophysics, and cosmology. For example, GPS systems must account for time dilation due to the high speeds of satellites relative to Earth.

8.3 Atomic and Nuclear Physics

The study of the atom's structure and the forces within the nucleus has led to significant advancements in understanding energy, radiation, and matter.

8.3.1 Atomic Structure

Atoms are the basic units of matter, consisting of a nucleus made of protons and neutrons, with orbiting around it.

Electron Orbitals: Electrons occupy specific orbitals or energy levels around the nucleus. When electrons move between orbitals, they absorb or emit energy in discrete amounts (quanta).

Photon Emission: The energy difference between levels determines the wavelength (color) of light emitted or absorbed, leading to the unique spectral lines of elements.

Applications: Atomic structure understanding is crucial in fields like chemistry, spectroscopy, and materials science.

8.3.2 Nuclear Physics

Nuclear physics studies the nucleus and interactions between protons and neutrons, which are bound by the ***strong nuclear force***. Key concepts include:

Radioactivity: Certain elements have unstable nuclei and release particles and energy as they decay into stable forms. This decay occurs via alpha, beta, or gamma radiation.

Nuclear Reactions: Involve changes in the nucleus, such as fusion (combining nuclei) and fission (splitting nuclei). Both release immense energy and have applications in power generation and nuclear weapons.

Applications of Nuclear Physics: Nuclear physics is central to energy generation in nuclear reactors, medical imaging (such as PET scans), and radiotherapy for cancer treatment.

8.4 The Standard Model of Particle Physics

The Standard Model is the most widely accepted theory describing the fundamental particles and forces in the universe (except for gravity).

Elementary Particles: The Standard Model categorizes particles into quarks, leptons, and bosons.

1. Quarks: Combine to form protons and neutrons.

2. Leptons: Include electrons and neutrinos.

3. Bosons: Particles that mediate forces. For instance, the photon mediates the electromagnetic force, and the Higgs boson is responsible for giving particles mass.

Forces in the Standard Model: The model describes three fundamental forces:

1. Electromagnetic Force: Acts between charged particles.

2. Weak Nuclear Force: Responsible for radioactive decay.

3. Strong Nuclear Force: Holds protons and neutrons together in the nucleus.

Applications: Particle physics research underpins technology in fields such as medical imaging and treatments, radiation therapy, and even computer chips.

8.5 Key Experiments in Modern Physics

Modern physics includes groundbreaking experiments that changed our understanding of reality:

8.5.1 Double-Slit Experiment

This experiment demonstrates wave-particle duality by showing that particles (like electrons) can behave like waves, creating an interference pattern when not observed, but behaving like particles when measured.

8.5.2 Photoelectric Effect

Einstein's work on the photoelectric effect showed that light can eject electrons from a material only if it has enough energy (frequency). This effect provided evidence for the particle nature of light and earned Einstein a Nobel Prize.

8.5.3 Large Hadron Collider (LHC)

The LHC is a particle accelerator that collides protons at high speeds to study fundamental particles. It confirmed the existence of the Higgs boson, a breakthrough in understanding mass and validating the Standard Model.

8.6 Modern Physics and Technology

The principles of modern physics have transformed many fields, leading to technology we use every day.

Semiconductors: Found in computer chips and electronic devices, they rely on quantum mechanics to control electron behavior.

Lasers: Used in everything from surgical instruments to communications and manufacturing.

Nuclear Power: Utilizes nuclear reactions to generate electricity, offering a powerful source of energy.

Quantum Computing: Uses principles of quantum mechanics, like superposition and entanglement, to perform complex calculations faster than classical computers.

Medical Imaging: Techniques such as MRI, CT, and PET scans rely on quantum mechanics and nuclear physics to create detailed images of the body.

8.7 Philosophical Implications of Modern Physics

Modern physics challenges our understanding of reality, time, and space, raising philosophical questions about the nature of existence.

Determinism vs. Probability: Classical physics is deterministic, but quantum mechanics introduces probability, suggesting that not all events are predictable.

Nature of Reality: Concepts like superposition and entanglement challenge traditional views of reality, suggesting that particles can exist in multiple states simultaneously or be instantaneously connected over distances.

Time and Space: Relativity redefines space and time as interconnected, showing that they can be warped by gravity and high speeds.

8.8 Summary

Modern physics offers a revolutionary perspective on nature, uncovering the workings of the atomic and subatomic world, as well as the fabric of space and time itself. Quantum mechanics reveals that matter and energy operate on probability and duality, while relativity reshapes our understanding of gravity, time, and space. By studying modern physics, we gain not only technological advancements but also profound insights into the universe's fundamental nature.

This journey from classical to modern physics represents humanity's quest to understand the mysteries of the universe, expanding the boundaries of what we know and what we believe to be possible.

Chapter 9: Simple Physics Experiments

Physics is best understood through observation and experimentation. Conducting simple experiments allows you to see core principles of physics in action, reinforcing theoretical concepts with tangible examples. This chapter presents easy-to-perform physics experiments that require minimal equipment and offer valuable insights into mechanics, thermodynamics, optics, electromagnetism, and more. Whether you're a student, teacher, or curious learner, these experiments will help you grasp physics in an interactive and fun way.

9.1 Safety Guidelines

Before conducting any experiment, follow these safety tips:

Read Instructions Carefully: Ensure you understand the steps before starting.

Use Appropriate Safety Gear: Wear safety goggles, gloves, or other protective equipment as needed.

Work in a Safe Environment: Conduct experiments on a stable surface, away from flammable materials, and in a well-ventilated area.

Supervise Younger Participants: Ensure children are supervised during experiments.

Clean Up Afterward: Dispose of materials properly and clean your workspace.

9.2 Mechanics Experiments

9.2.1 Measuring Acceleration Due to Gravity

Purpose: To calculate, the acceleration due to gravity.

Materials: A stopwatch, a measuring tape, and a small object (like a ball).

Procedure:

1. Measure a vertical height (h) from which the ball will be dropped.

2. Drop the ball from rest and measure the time (t) it takes to hit the ground.

3. Use the formula: $g = \frac{2h}{t^2}$

Concepts Illustrated: Free fall, acceleration, and kinematics.

9.2.2 Building a Simple Pendulum

Purpose: To explore the relationship between pendulum length and period.

Materials: String, a small weight, and a stopwatch.

Procedure:

1. Attach the weight to the string and suspend it from a fixed point.

2. Pull the weight to one side and release it gently, letting it swing.

3. Measure the time it takes for the pendulum to complete one oscillation (the period).

4. Repeat with strings of different lengths and note the changes in period.

Concepts Illustrated: Simple harmonic motion, periodic motion, and gravitational effects.

9.3 Thermodynamics Experiments

9.3.1 Observing Heat Transfer

Purpose: To demonstrate conduction, convection, and radiation.

Materials: Metal rod, hot water, and a thermometer.

Procedure:

1. Place one end of the metal rod in hot water and observe how the heat transfers along the rod (conduction).

2. Hold your hand above the water and feel the heat rising (convection).

3. Hold your hand to the side of the container and feel the heat radiating outward (radiation).

Concepts Illustrated: Heat transfer modes, thermal conductivity.

9.3.2 Creating a Homemade Thermometer

Purpose: To observe thermal expansion.

Materials: A clear plastic bottle, water, rubbing alcohol, a straw, and food coloring.

Procedure:

1. Mix equal parts of water and rubbing alcohol, adding a few drops of food coloring for visibility.

2. Fill the bottle with the mixture and insert the straw, sealing the top with clay to make it airtight.

3. Warm the bottle with your hands or place it in sunlight and observe the liquid rising in the straw.

Concepts Illustrated: Thermal expansion, temperature changes.

9.4 Electromagnetism Experiments

9.4.1 Making a Simple Electromagnet

Purpose: To demonstrate how electricity produces magnetism.

Materials: A battery, insulated copper wire, and an iron nail.

Procedure:

1. Wrap the copper wire tightly around the nail, leaving two ends free.

2. Connect the wire ends to the battery terminals.

3. Use the nail to pick up small objects like paper clips.

Concepts Illustrated: Electromagnetic fields, current, and magnetism.

9.4.2 Generating Electricity with a Lemon

Purpose: To create a simple chemical cell.

Materials: A lemon, a copper coin, a zinc nail, and wires with alligator clips.

Procedure:

1. Insert the copper coin and zinc nail into the lemon, keeping them apart.

2. Connect wires to the metal pieces and attach them to a small LED or multimeter.

3. Observe the LED light up or the current reading on the multimeter.

Concepts Illustrated: Electrochemistry, voltage generation.

9.5 Optics Experiments

9.5.1 Observing Refraction with a Glass of Water

Purpose: To demonstrate the bending of light.

Materials: A clear glass, water, and a pencil.

Procedure:

1. Fill the glass with water and place the pencil inside.

2. Look at the pencil from the side and observe how it appears bent at the water's surface.

Concepts Illustrated: Refraction, optical density.

9.5.2 Creating a Simple Lens

Purpose: To demonstrate light focusing.

Materials: A clear plastic bag, water, and a flashlight.

Procedure:

1. Fill the plastic bag with water and tie it to create a convex shape.

2. Shine the flashlight through the water-filled bag onto a wall or paper and observe the focused light.

Concepts Illustrated: Lens properties, focusing of light.

9.6 Waves Experiments

9.6.1 Observing Sound Waves

Purpose: To visualize vibrations and sound.

Materials: A metal pan, plastic wrap, rice, and a speaker.

Procedure:

1. Cover the pan with plastic wrap and sprinkle rice on top.

2. Place the pan near the speaker and play loud music or tones.

3. Observe the rice jumping due to sound vibrations.

Concepts Illustrated: Sound waves, vibration, and resonance.

9.6.2 Creating a Ripple Tank

Purpose: To study wave properties like reflection and interference.

Materials: A shallow tray, water, and a small object (like a pen cap).

Procedure:

1. Fill the tray with water and let it settle.

2. Gently tap the water with the object to create ripples.

3. Observe how the ripples interact with the edges of the tray and each other.

Concepts Illustrated: Wave behavior, interference, and reflection.

9.7 Summary

Simple physics experiments bring theories to life, making learning interactive and engaging. From understanding how pendulums swing to generating electricity with lemons, each experiment showcases fundamental principles in a hands-on way. Conducting these experiments not only reinforces theoretical knowledge but also fosters curiosity and critical thinking, laying the groundwork for further exploration in the world of physics.

Chapter 10: Applying Physics in Everyday Life

Physics is not just an academic subject confined to labs or classrooms, it is all around us. From the moment you wake up to the moment you sleep, physics is at play in countless ways, influencing the technology you use, the environment you experience, and even your own body. This chapter explores the practical applications of physics in everyday life, highlighting its importance in areas such as transportation, communication, healthcare, energy, and even the simple act of walking.

10.1 Physics in Transportation

10.1.1 The Science of Motion

Physics governs how vehicles move. Concepts such as force, friction, and acceleration ensure smooth and efficient travel.

Newton's Laws of Motion: Cars accelerate (Newton's Second Law), maintain momentum (Newton's First Law), and decelerate due to brakes (Newton's Third Law).

Friction and Traction: Tires use friction to grip the road, enabling safe turns and stops. Understanding friction helps design better tires and road surfaces.

Aerodynamics: The streamlined shapes of cars and airplanes reduce air resistance, improving fuel efficiency and speed.

10.1.2 Safety Features

Physics is crucial in designing safety mechanisms such as:

Seat Belts: Use inertia to restrain passengers during sudden stops.

Airbags: Deploy to reduce the impact force during collisions by increasing the time over which the force acts.

Crumple Zones: Absorb and dissipate energy from impacts to protect passengers.

10.2 Physics in Communication Technology

10.2.1 Wireless Communication

Electromagnetic Waves: Technologies like Wi-Fi, mobile phones, and radio broadcasts rely on the transmission and reception of electromagnetic waves.

Fiber Optics: High-speed internet and telecommunications use light to transmit data through optical fibers, which operate on the principle of total internal reflection.

10.2.2 Sound and Microphones

Microphones and Speakers: Convert sound waves into electrical signals and vice versa, enabling communication and entertainment systems.

10.3 Physics in Healthcare

10.3.1 Medical Imaging

Physics has revolutionized diagnostic methods:

X-Rays: Use high-energy electromagnetic waves to image bones and tissues.

MRI (Magnetic Resonance Imaging): Utilizes magnetic fields and radio waves to create detailed images of internal organs.

Ultrasound: Uses sound waves to image soft tissues and monitor fetal development.

10.3.2 Lasers in Medicine

Lasers, which are based on principles of light amplification, are used in:

Eye surgeries like LASIK.

Precise cutting in surgical procedures.

Cancer treatments through photodynamic therapy.

10.4 Physics in Energy and Power

10.4.1 Electricity in Daily Life

Ohm's Law: Governs how electricity flows through devices in your home, like lights, fans, and appliances.

Electric Generators: Convert mechanical energy into electrical energy using electromagnetic induction, supplying power to homes and industries.

10.4.2 Renewable Energy Sources

Solar Panels: Convert sunlight into electricity using the photovoltaic effect.

Wind Turbines: Use principles of rotational motion and aerodynamics to generate electricity from wind energy.

Hydropower: Converts the potential energy of water into electricity using turbines.

10.5 Physics in Everyday Gadgets

10.5.1 Smartphones

Smartphones integrate several physics concepts:

Touch Screens: Capacitive screens detect changes in electric fields when you touch them.

Gyroscopes and Accelerometers: Detect motion and orientation, enabling features like auto-rotation and motion tracking in games.

10.5.2 Refrigerators and Air Conditioners

Use principles of thermodynamics, including heat transfer and the refrigeration cycle, to cool spaces or preserve food.

10.5.3 LED Lights

Rely on electroluminescence, where electrons release energy as light when passing through a semiconductor.

10.6 Physics in Sports and Fitness

10.6.1 Motion and Force in Sports

Kinematics: Explains how a soccer ball curves or a gymnast rotates.

Projectile Motion: Governs the trajectory of a basketball shot or a javelin throw.

Friction: Provides grip in sports like tennis or skiing.

10.6.2 Wearable Technology

Fitness trackers use physics concepts like acceleration and heart rate monitoring to measure physical activity.

10.7 Physics in Household Activities

10.7.1 Cooking

Heat Transfer: Cooking involves conduction (pan to food), convection (oven or boiling water), and radiation (microwave or grill).

Pressure Cookers: Use the principles of pressure and boiling point elevation to cook food faster.

10.7.2 Washing Machines

Operate using centrifugal force to remove water from clothes during the spin cycle.

10.8 Physics and the Environment

10.8.1 Understanding Weather

Thermodynamics: Explains temperature variations, wind patterns, and the formation of storms.

Water Cycle: Driven by the energy from the sun, involving evaporation, condensation, and precipitation.

10.8.2 Climate Change

Greenhouse Effect: Involves the absorption and re-emission of infrared radiation by gases like carbon dioxide, leading to global warming.

10.9 Physics and Human Movement

10.9.1 Walking and Running

Biomechanics: Explains how forces act on your body during movement.

Center of Gravity: Determines balance and stability.

10.9.2 Hearing and Vision

Sound Waves: Enter your ears and are processed by your brain.

Light Waves: Interact with the lens and retina in your eye to produce vision.

10.10 Summary

Physics is deeply intertwined with our daily lives, from the transportation we use and the gadgets we rely on, to the natural phenomena we observe and the healthcare advancements we benefit from. Understanding these applications not only helps us appreciate the science behind everyday occurrences but also empowers us to make informed decisions and innovations. Physics bridges the gap between theory and practice, proving itself indispensable to modern living.

Conclusion

Physics is the cornerstone of our understanding of the universe, shaping the way we perceive and interact with the world around us. From the smallest particles to the vastness of space, it provides answers to fundamental questions and equips us with the tools to innovate, solve problems, and improve our lives.

In this **Physics Foundations: Your First Guide to the Universe**, we began by exploring the **basics of physics**, uncovering its significance as a scientific discipline. We then delved into the **core concepts**, such as motion, energy, and forces, which form the foundation of the subject. With an understanding of essential **tools and resources**, we highlighted how to approach physics effectively, whether through experimentation, calculations, or using modern technologies like simulations.

We examined specialized topics like **mechanics, thermodynamics, electromagnetism, waves and optics, and modern physics**, each opening doors to deeper insights and real-world applications. Through practical experiments, we saw how physics comes alive, bridging theory and practice. Finally, we explored the myriad ways physics applies to daily life, revealing its role in transportation, healthcare, communication, and the environment.

Key Takeaways

1. **Physics is Universal:** It explains the forces, interactions, and principles governing everything from atoms to galaxies.

2. **Practical Applications are Everywhere:** Physics enhances technology, healthcare, energy production, and even our everyday decisions.

3. **Curiosity Fuels Learning:** The experiments and activities shared are a starting point for your own explorations.

4. **Critical Thinking is Key:** Understanding physics hones your ability to analyze, predict, and solve problems logically.

Looking Ahead

This first guide is just the beginning of your journey into physics. Whether you're a student, educator, or enthusiast, the principles and practices shared here can serve as a springboard for deeper exploration. As you continue learning, remember that physics is not just about equations and theories, it's a way of seeing the world, a path to innovation, and a lens through which we uncover the mysteries of existence.

Embrace curiosity, ask questions, and let the principles of physics guide you toward a deeper understanding of the universe. Thank you for embarking on this journey with the **Physics Foundations: Your First Guide to the Universe**, your adventure into the world of physics has just begun!

Appendices

The appendices provide additional resources, formulas, reference materials, and tools to complement the content of this Physics first Guide. These sections are designed to serve as quick references for readers as they explore the world of physics.

Appendix A: Key Physics Formulas

*Mechanics

1. Newton's Second Law:

$$F = ma$$

2. Kinematic Equations (for constant acceleration):

$$v = u + at$$

$$s = ut + \frac{1}{2}at^2$$

$$v^2 = u^2 + 2as$$

Where u is initial velocity, v is final velocity, a is acceleration, s is displacement, and t is time.

3. Work-Energy Principle:

$$W = F \cdot d \cdot \cos\theta$$

4. Potential Energy:

$$PE = mgh$$

5. Kinetic Energy:

$$KE = \frac{1}{2}mv^2$$

*Thermodynamics

1. First Law of Thermodynamics:

$$\Delta U = Q - W$$

2. Efficiency of a Heat Engine:

$$\eta = \frac{W_{output}}{Q_{input}} \cdot 100\%$$

*Electromagnetism

1. Ohm's Law:

$V = IR$

2. Coulomb's Law:

$F = k_e \frac{q_1 q_2}{r^2}$

3. Faraday's Law of Induction:

$Emf = -N \frac{d\phi_B}{dt}$

*Waves and Optics

1. Wave Speed:

$v = f\lambda$

2. Lens Formula:

$\frac{1}{f} = \frac{1}{v} - \frac{1}{u}$

3. Snell's Law:

$n_1 sin\theta_1 = n_2 sin\theta_2$

*Modern Physics

1. Einstein's Energy-Mass Relation:

$E = mc^2$

Appendix B: Units and Constants

Fundamental Units

Quantity	Unit	Symbol	SI Unit name
Length	meter	m	Meter
Mass	kilogram	kg	Kilogram

Time	second	s	Second
Electric current	ampere	A	Ampere
Temperature	kelvin	K	Kelvin
Amount of substance	mole	mol	Mole

Physical Constants

Constant	Value	Units
Speed of light (c)	3.00×10^8	m/s
Gravitational constant (G)	$6.674 * 10^{-11}$	$N.m^2/kg^2$
Planck's constant (h)	$6.626 * 10^{-34}$	$J.s$
Elementary charge (e)	$1.602 * 10^{-19}$	C

Appendix C: Experiment Checklist

Basic Tools

Measuring tape or ruler

Stopwatch

Spring balance

Multimeter

Thermometer

Magnifying lens or basic optical equipment

Materials for Experiments

String and weights (e.g., pendulums)

Small objects for free-fall experiments

Batteries, wires, and nails (for electromagnetism)

Transparent containers and liquids (for optics and refraction experiments)

Appendix D: Glossary of Key Terms

The appendices provide the necessary tools and resources to make this first guide more functional for students, teachers, and enthusiasts alike. Use them as quick references as you deepen your understanding of physics!

Acknowledgments

The author of the book: Yannick Mbambu and his support team

As a citizen of the world, he is a writer, then a student and researcher in math-physics.